YOUR KNOWLEDGE HAS VALUE

- We will publish your bachelor's and
  master's thesis, essays and papers

- Your own eBook and book -
  sold worldwide in all relevant shops

- Earn money with each sale

Upload your text at www.GRIN.com
and publish for free

Bibliographic information published by the German National Library:

The German National Library lists this publication in the National Bibliography;
detailed bibliographic data are available on the Internet at http://dnb.dnb.de .

Imprint:

Copyright © 1990 GRIN Verlag
Print and binding: Books on Demand GmbH, Norderstedt Germany
ISBN: 9783668748910

This book at GRIN:

https://www.grin.com/document/432632

Dieter Wessels

Seismic Measurements Of Waves In Seams As A Geophysical Tool In The Extraction Of Raw Materials

GRIN Verlag

Seismic Measurements Of Waves In Seams As A Geophysical Tool In The Extraction Of Raw Materials

by Dieter Wessels, M.E.

1 Introduction

German studies have found that a significant proportion, between 16% and 20%, of all the raw materials deposited in the form of fluff are geologically disturbed. /1/ [1]

Due to the extreme mechanization of the mining companies in the German coal industry (approx. 99.3% in 1980) and the associated capital tie-up in the technology, geological disturbances in the seams, which may hinder or bring to a standstill, adversely affect the production results. So it costs depending on the nature of the fault and the longwall equipment 300,000 DM up to 2.5 million DM to pass through a fault in the seam. If even a mining operation has to be abandoned because it can not be passed through, the costs increase by almost 3 times as much as production, rental of machinery, construction of a new facility and relocation. / 2 /

From the above figures, it immediately becomes clear how important it is to explore in advance the homogeneity of the raw material in the seam of planned mining operations.

There are various methods available for exploring the apron, which have different levels of security and statements.

As one of the safest methods of forecasting of underground situations, seismic measurements of waves in seams have emerged, which are to be dealt with in this work.

---

[1] The given numbers refer to the numerical sequence in the bibliography as reference

The work is an interdisciplinary bridge between geophysics and practical application in the extraction of flooded raw materials; In the present specification and introduction coal is chosen because of its importance to the economy.

The basic system is for all seamy ones Deposits are equal and of general importance. [2]

---

2   Seam, mining name for the accumulation of sedimentary formed, usable minerals in the form of a layer that has in relation to their thickness a large length and width and is limited by almost parallel surfaces" from: The Little Mining Dictionary, Verlag Glückauf GmbH, Essen, 1983. In the following, explanations of mining technical terms are taken from all of this cited source, without this being stated again.

The prerequisite for the application of the Flözwellenseismik is the presence of a seismic channel, in which there are between the above and below the seam layers and himself sufficient density differences. An example is a coal seam between two rock layers.

The representation is given because the density contrast between coal and adjacent rock is approximately 1: 1.75 to 2, and the coal is a medium in which seismic waves propagate more slowly than in the surrounding rock. Basically, the wave velocity increases with the density of the medium.

The ratio of the propagation velocities of the waves is equal to that of the density contrast. The emergence of Flözwellen is systematically since r provided in Figure 1.

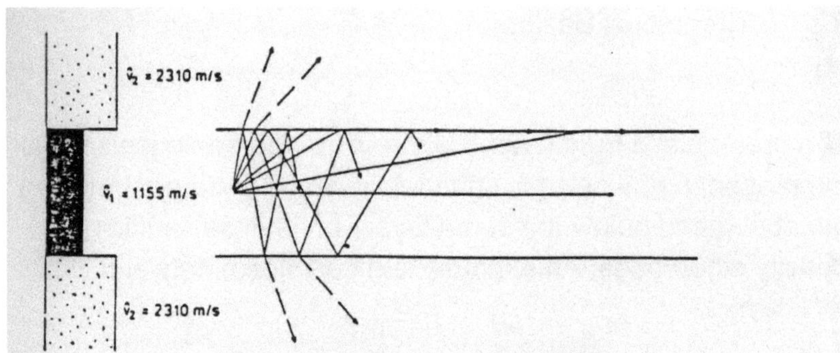

Figure 1: Formation of seam waves by mining by reflection of seismic waves at the interfaces of a seam / 13 /

Seismic waves, which are excited in the seam, can not radiate spherical, as it really is their nature, but are partially reflected at the two interfaces to the hanging and lying and thus form a special wave, the "seam-wave". These waves are excited in the seam, guided through the seam and received in the seam again.

Meet these waves on a fault [3], which cuts the sea this technique is based on the measuring method of the seismic of seam-waves in such a way that behind the interface rocks of different density than coal are present, they are reflected. If there is no disturbance, the flute waves spread unhindered.

---

[3] A geological change of the original form of deposition of a mineral or rock. Disturbances can be caused by tectonic forces, which manifest themselves in the form of layer tears (distortions).

This technique is based on the measuring method of the seam seismic/ 7 /.

To describe seam-waves in more detail, it requires the following differentiation:

1. compression waves	- P waves
2. shear waves	- S waves
a) vertically polarized	- SV waves
b) horizontally polarized	- SH waves

Depending on the geometry and particle movement, flute waves are differentiated into those of the Rayleigh-type and the Love-type. The Rayleigh type consists of P- and SV-waves, those of the Love-type exclusively of SH-waves.

Symmetrical waves of the Love type have proven to be well-suited for measurement practice, and can be easily separated from other types of waves by targeted orientation of the vibration pick-ups / 10 /. Like all guided waves, flute waves have a dependence between the propagation velocity and the wavelength, they disperse [4]. In the first place, the dispersion is

---

4   In physics, dispersion is defined as the dependence of wave properties on wavelength.

bound to the different shear wave velocities of coal and secondary rock, but it also depends on the channel width, means seam height.

The effects of the dispersion are in detail:

- With increasing dispersion, the energy of the waves is more strongly bound to the soft medium, so that at higher frequency, the wave can not leave the medium.

- Even with an excitation of the wave over short temporal influence of the pulse, the flute wave appears as a long, time-extended wave trains due to the dispersion; this causes a differentiation between the faster phase velocity and the slower group velocity. The relationship is shown in Figure 2:

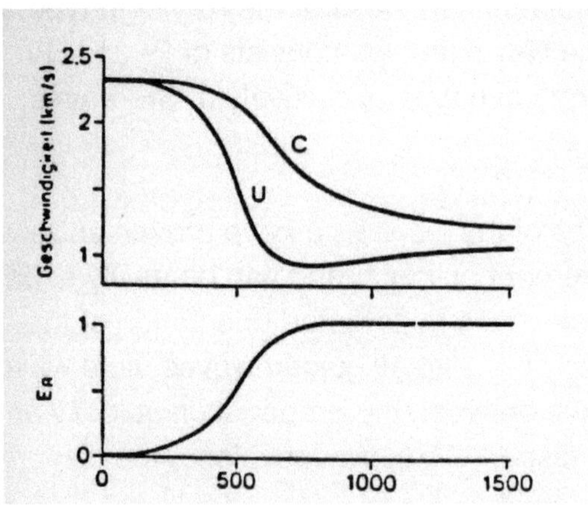

Fig. 2: Dispersion curve of the symmetrical flute waves of the Love-type / 10 /.

with: C = phase velocity

U = group speed

E R = ratio of energy in the seam to total
power

x - axis = frequency [Hz] * Seaminess [m]

The group velocity curve has a pronounced minimum
at which the kinetic energy carried in the medium is
nearly equal to the total energy. This minimum is
called the Airy phase, which stands out at high
frequencies of almost 500 Hertz.

- The seam-wave is the longer by the dispersion, the
greater the distance between the excitation and
reception.

The Airy phase is used in subsurface seismic as a useful
signal, while all other wave types represent noise that can be
separated as low frequency by filtering. The principle is shown
visually clearly in Figure 3:

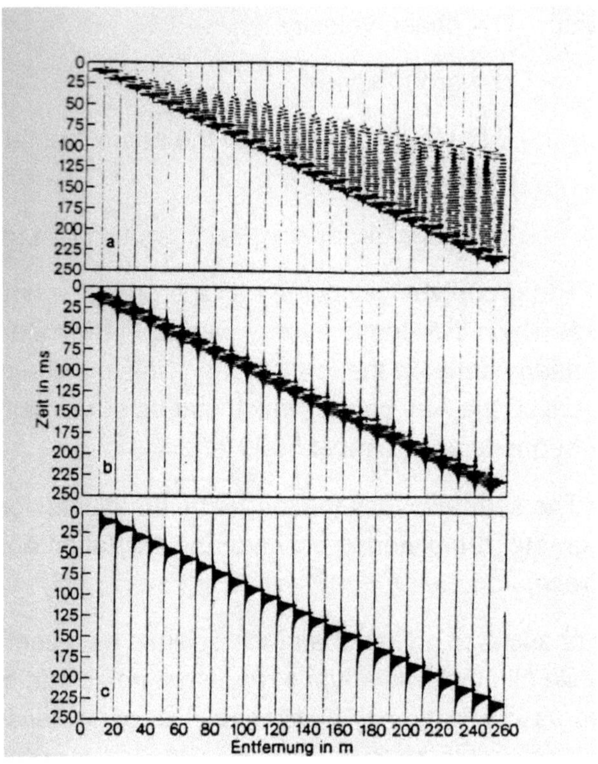

Fig. 3: Theoretical Flözwellenseismogramme / 6 /

with: a) Registered Flözwellenzug

b) filtered out Airy phase

c) calculated envelope [5] the Airy phase

---

5   mathematically a curve that envelops a group of curves

The seismogram time in part a) the flooding of the flooding waves depending on the way, and at the end of the trains the Airy phase is easily recognizable. Part b) clearly shows the separated Airy phase by filtering. Since the digital processing of the pure Airy phase is only possible through the formation of an enveloping envelope in order to be able to neglect the phase shift in the filtering, it is necessary to dispense with a part of the information obtained in the seismogram. The envelopes are reproduced separated in part c) / 2 /, / 3 / and / 9 /.

3. Measuring Principles

To explore the apron by seam-waves two methods are available. The reflection measurement makes it possible to detect tectonic disturbances locally. The measurement of the sound transmission provides data for the determination of interference contributions in comparison to the seam thickness, which means crosstalk disturbances are detected. For the accuracy, it proves useful to apply both procedures in parallel.

3.1   Reflection Measurements

An explosion at a depth of about 2 meters causes an energetic stimulus, which causes waves to fan out. These are reflected according to the laws of optics, when they encounter a problem, because behind the seam transfer are other physical constants exist as before. The geophone introduced in the seam [6] pick up the reflected waves and pass analogue voltage signals to a recording device, where they are recorded in the form of seismograms, as shown in Figure 4.

---

6   A geophone is an electro-mechanical transducer that converts vibrations in media into analogue voltage signals. They are used in mining for exploration.

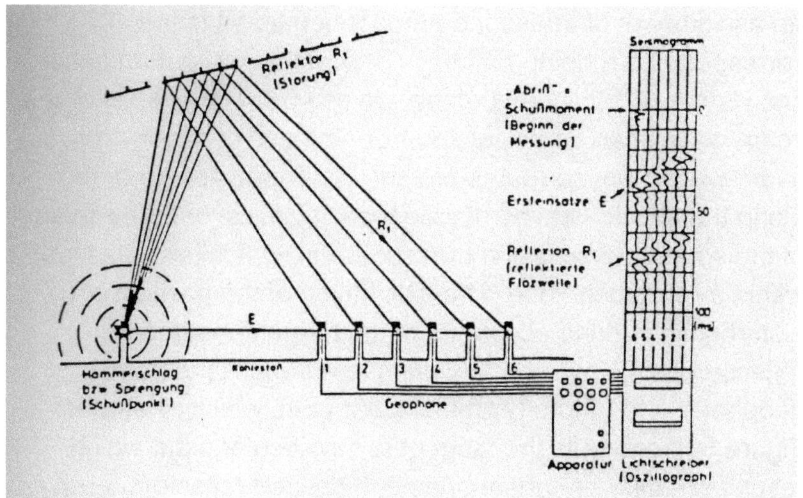

Fig. 4: Scheme of a reflection measurement / 11 /

In order to increase the accuracy, the method of multiple overlapping has proven itself in the case of underground measurements. Each reflection point of the disturbance is recorded multiple times from different firing and geophone distances / 8 /. The data belonging to a reflection point is added (stacked) after applying a dynamic correction. The dynamic correction reflection time differences are eliminated and the various weft and geophone distances reduced to the value zero. By means of the stacking one obtains an amplification of the in-phase reflection signals and a simultaneous attenuation of the interference signals. Increasing the coverage of the subsoil points further improves the quality of the information / 13 /.

An assignment of the sound propagation times to the corresponding running paths is not possible without an exact knowledge of the speeds of the waves / 6 /. The speeds of the seam-waves can be calculated from the use of directly run seam-waves whose routes are known. These are either the along the shock [7] waves of reflection seismics, or flute waves of the sound transmission measurement, which have passed within a relaxation zone. The calculation of the position of a disturbance is done according to the tangent method. The reflection points, which have the same transit time from a geophone, are naturally on a circular path, which illustrates Figure 5. If one puts the tangent to the circular path, which result from different geophones, the desired reflector corresponds to the common circular tangent /4/.

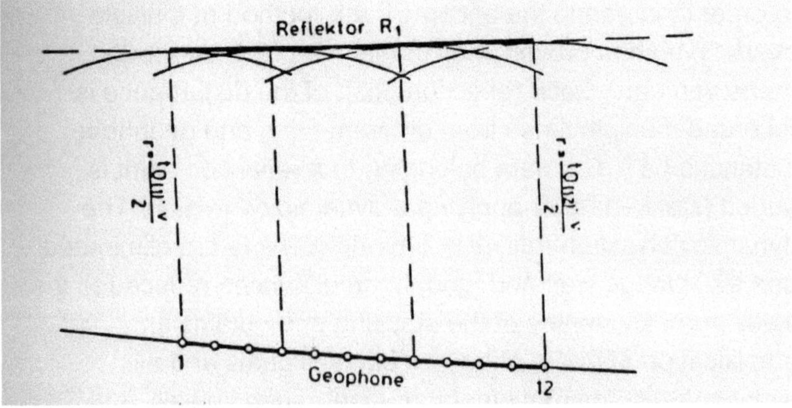

Figure 5: Principle of the tangent method / 3 /

---

7   Stoss: Lateral boundary surface of a cavity

The range of a reflection measurement is usually 100 times the seam thickness, in exceptional cases distances up to 200 times the seam thickness have been bridged /2/ .

3.2   Through-Transmission Measurement

The through-transmission measurement requires two mines, which are arranged inclined to each other. As outcrops can be used a longwall or a track for the stationing of geophones and another track for energy stimulation. Another possibility is to trigger the energy stimulation in a deep hole in the area of the seam and to register in a section of the pit building. The range of a transmission measurement is limited to 1000 times the seam thickness and suitable deep holes are present in only very rare cases. / 5 /

Figure 6 shows the principle of sound transmission.

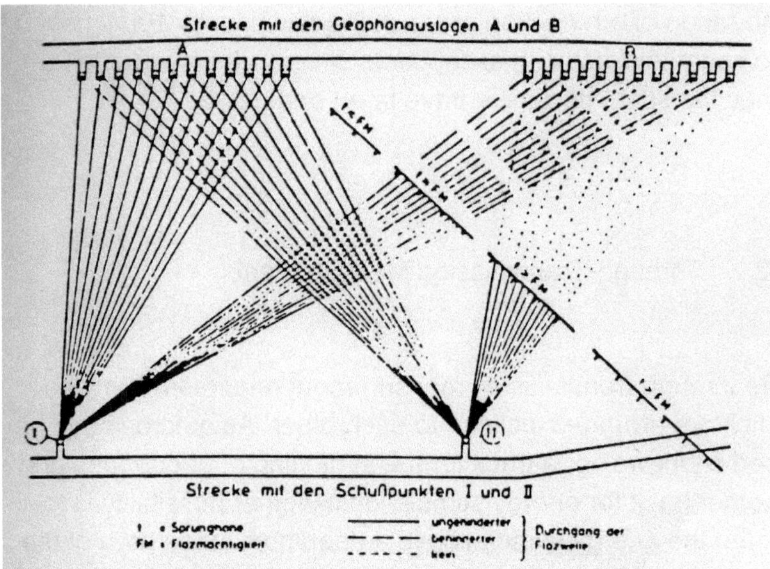

Figure 6: Scheme of a transmission measurement / 11 /

Since the Airy phase is bound to the seam, the propagation of this wavefront is hindered by tectonic perturbations in the direction of radiation. If there is a fault, the jump amount exceeds the seam thickness, the reception of seam-waves is not possible. If clear flute waves are received during the sound transmission, this indicates an uninterrupted wave channel.

The numerous steps that lie between a clear reception and the absence of seam-waves are difficult to interpret. An unambiguous assignment to the jump height of a non-zero offset disturbance is not possible. However, in case of a time delay of seam-wave on the diffraction at an obstacle are assumed / 3 /.

In order to be able to separate disturbing compression waves from the Love-type shear waves, two-component geophones are used. While the X component is oriented parallel to the seam and parallel to the joint, the Y component can oscillate parallel to the seam and perpendicular to the joint. The component directed to the shot point preferably registers the compression waves, the other measures the Love-type shear wave. However, since the geophones do not point to the shot point and both components register both compression and shear wave components, the geophone axes are computationally "rotated" during data processing /2/ /9/.

4 Measuring Equipment And Implementation Of The Measurements

After the completion of the impact-weather-protected digital data acquisition system MDH-01 of the company Texas Instruments, a breakthrough in the field of underground seismic was achieved in 1978. This system corresponds to the above-used equipment DSF V.

The previously used analogue equipment provided results with insufficient accuracy, so that by later Auffahrungen only about 2/3 of the reflection and 80% of the sound transmission results could be confirmed /11/ /4/.

The digital measuring equipment consists of two analogue modules with amplifiers, multiplexers and analog-to-digital converters, a control unit that controls the processes within the equipment and a magnetic tape module that stores the incoming data. The storage takes place after the analogue oscillations have been sampled at time intervals (0.5 ms) and decomposed into discrete numerical values.

A light beam oscilloscope is used to pick up certain geophone groups using an input array and for continuous monitoring of the measurement results. The oscilloscope records the vibrations of 12 geophones on photo paper, which develops itself. The shielding gas control unit is used to monitor fired-weather protection. The modules are connected to 12-volt nickel-cadmium batteries and are cabled together. The individual units are connected via hoses to the protective gas control unit, which is connected to a nitrogen cylinder with a reduction. The flushing of the system with 2.5 m 3 of the inert gas takes 45 minutes. The inert gas control unit unlocks the main switch after the corresponding flushing time. Falling below a minimum pressure or exceeding a maximum value triggers the main fuse. Leakage losses that can occur due to the slight overpressure in the system are compensated by the shielding gas control unit by opening a valve of /2/ /12/. A sketch of the device can be found in Fig. 7.

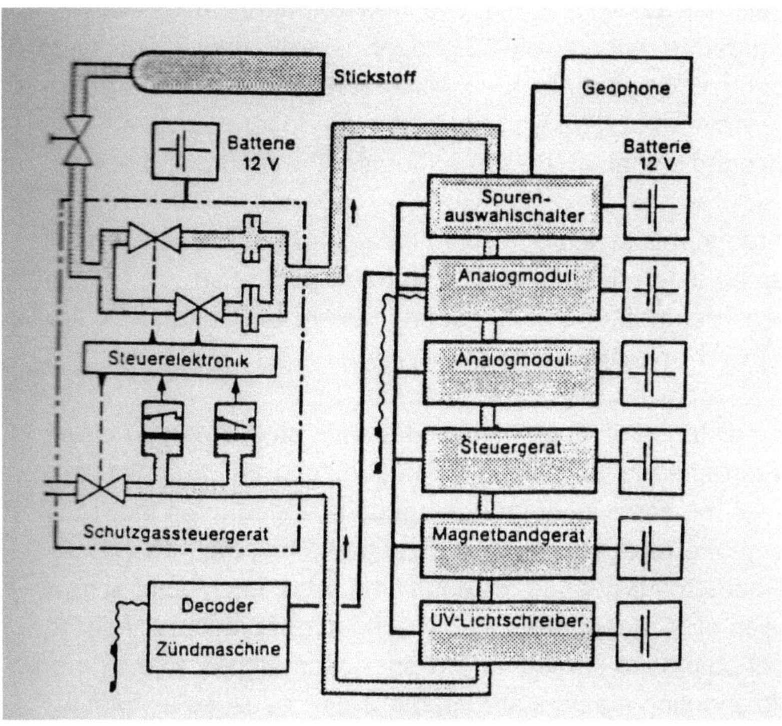

Fig. 7: Impact-resistant measuring equipment / 6 /

48 individual components (24 geophones) can be connected
to the apparatus, of which the input array field picks out 24 for
each measurement. The two-component geophones are
installed in 2 m deep drilled holes in the measuring section
and firmly coupled by pumping a rubber bellows. By
incorporating the geophones, absorption of the flute waves
and the loosening zone at the impact can be largely excluded.
The geophones are made of non-sparking material; the rubber
parts are electrically conductive. The connection of the
geophones to the equipment is made by special PVC
sheathed cables (extensions and measuring cables). Each test
cable has 12 taps, so that two measurement cables are
required to respectively 12 geophones with their X and Y - to
connect components, the ignition of the explosive charge to
the wave excitation is triggered by the apparatus via an
encoded signal; at the same moment the recording begins.

Through this synchronization, the sound propagation times
can be determined exactly. For the transmission
measurement, up to 120 geophones are positioned at intervals
of 2 or 4 m in the drill holes.

The distance of the drill holes depends on the length of the
track to be measured and the desired degree of resolution.
Since the capacity of the apparatus is limited to 24 individual
components at a working interval of 0.5 ms, the transmission
is made to 12 geophones each time. After each blast, a new
group of 12 geophones will be connected. Since the individual
shot points are sometimes far apart, sometimes 1 km or more,
two shooting masters are used in such cases, so as to avoid

the time-consuming implementation of the unwieldy cable firing system.

For geophysics, the geophone holes can be used for detonation because both blast and geophone holes are in the same distance. Thus, depending on the start (distance of the shot point from the first geophone) only 6 or 12 shooting holes have to be drilled. The start depends on the course of the suspected disorder. As a rule, the reflection measurement is preceded by a transmission measurement, so that after the sound has been transmitted, the reflection on the already installed and wired geophones can be started immediately. After each blast, the measurement technician picks up a new geophone with the input array and simultaneously shifts another one out. The geophones 1 to 12 thus record the sound waves generated by the first blast, the geophones 2 to 13 the blast from shot hole 2, etc.

The geophones that are no longer needed for measurement are removed and re-stationed at the end of the display. Thus, with reflection seismics, regardless of the number of geophones, an arbitrarily long distance can be detected, this continuous construction and removal method is also conceivable for the sound transmission. However, in this case, the geophones would have to be transported back for the subsequent reflection measurement and reinstalled. This procedure requires too much time in a combined measurement.

The energy stimulation to produce the seam-wave done with small amounts of the explosive Weather Devinit III. The amount of charge is usually in the reflection seismic a cartridge (125 g) and 1 to 3 cartridges in the transmission. At a detonation rate of about 1700 m / s and the use of small amounts of charge, the explosive primarily causes the seam to vibrate, while the energy output to the neighbouring mountains is low. It is therefore in the interest of the process to work with the smallest possible charge quantities. The detonators used are special seismic detonator igniters with a delay of less than 1 ms / 2 /.

In order to improve the excitation of the seam in the frequency range of the Airy phase and of the horizontally polarized type, experiments have recently been carried out with well-dosable detonating cord. Other considerations are to generate the vibration energy mechanically by air pulser or vibrators, not least to achieve independence from the shooting. After all, per double layer up to 200 blasts / 10 / / 11 /.

The entire underground equipment has a weight of about 2.8 t. For underground transport, 7 containers are required /12/.

5 Susceptibility To Interference Of The Measuring System

The entire measuring system can be subdivided into two areas, the electronics section (equipment) with accessories and the measuring display. Despite constant maintenance, specific errors often occur in both areas.

5.1 Apparatus

The introduction of digital measurement technology with an impact-protected equipment has significantly improved the quality of the measurement data and its processing. However, this increase in quality brought with it an increased susceptibility to failure of the system. The equipment, which is housed in 7 boxes, is repeatedly exposed to heavy vibrations during transport underground. The impacts sometimes cause the circuit boards (electronic inserts) to jump out of their holders. By further shocks individual electronic elements can tear off or hairline cracks occur. The high susceptibility to interference of the electronic devices is problematic in two ways. On the one hand, the costs increase, depending on the duration of the repair work up to 1200 DM per hour. Second, seismic measurements require as few interfering signals as possible over a wide range.

Since the measurements can only be carried out during the standstill, this means that only the production-free weekends are available for cost reasons in an ongoing mining operation. The measuring duration is limited to this period. Repairs to the equipment thus lead to a shortening of the measurement program.

5.2   Measurement Display

The picking up of vibrations takes place via geophones, which are introduced at defined intervals in boreholes. For practical reasons, it is not always possible to drill holes in certain areas, such as electrical supply equipment, belt drives, etc. Furthermore, individual geophones fail due to contact errors.

6 Measurement Results For The Detection Of Faults And Measurements With Special Objectives

The testing of the digital measuring system was started with simple and known geological conditions. Under "simple" are to be understood as jumps or leaves with almost in falling in vertically ideas, "known" means that the location to be localized disorder by mining information after strike, incidence and rate of transfer are occupied.

These measurements were necessary to be able to test and optimize measuring methods, data processing and evaluation. The next step was to test the measurement system on simple, projected tectonics. Simple-projected disturbances are understood to mean those which, according to their strokes and incursions, assuming that both do not change (plane disturbance surface), can be projected into the hanging, lying, or unopened areas of the same seam.

Such projections often serve as planning material for further mining, although they sometimes involve major uncertainties. The test of the digital measuring method in both cases described was very successful. Thus, all mining and drilling tests carried out in this context showed good agreement with the floe-seismic predictions.

In the next step, the digital measurement method has been tested on complicated interferences of various kinds:

- Thrusting: They differ from branching seismic jumps by small angles of inclination between fault and seam (10gon - 40gon),

- Disturbance bundle: several disturbances in close succession, whereby the individual amounts of discord may be smaller and larger than the filament size,

- Leaching: Syngeneically sedimentary floe regularity (erosion gullies).

The investigations carried out on complicated geological-tectonic conditions did not give satisfactory results, which was not to be expected with the degree of difficulty of the manifold manifestations.

The desired extension of the penetration depth could be realized in the reflection process. Before the beginning of the investigations, the maximum value at 100 times the seam thickness was 160 times that. A value about 200 times is certainly achievable, but was not detectable by measurements.

In cases where disturbances were to be located at greater distances, unexpectedly closer preliminary disturbances occurred, which blocked a deeper penetration of the floe waves. The theoretical penetration depth for penetrating measurements of 1000 times the seam thickness could not be reached so far. Widths up to 440 times the seam thickness could be registered.

The statistical evaluation of the measurements carried out so far, or their results and reviews, is as follows: (as at 31.12.1989)

BAG Niederrhein	50 measurements	45% tested
BAG lip	35	32%
BAG Westfalen	14	13%
Sophia Jakoba, AV, Saarbergwerke	12	10%

The following list distinguishes by tectonic characteristics and by verification results:

<u>A total of:</u>

Measurements taken:	111
Mining checked:	65 = 59%
confirmed by:	59 = 91%
not confirmed:	6 = 9%
not verifiable:	7 = 6%
still to check:	39 = 35%

<u>simple tectonics</u>

Measurements taken:	26
Mining checked:	19 = 73%
confirmed by:	19 = 100%
not confirmed:	0
not verifiable:	3 = 12%
still to check:	4 = 15%

<u>complicated tectonics</u>

Measurements taken:	75
Mining checked:	40 = 54%
confirmed by:	35 = 88%
not confirmed:	5 = 12%
not verifiable:	4 = 5%
still to check:	31 = 41%

<u>heights</u>

Measurements taken:	10
Mining checked:	6 = 60%
confirmed by:	5 = 83%
not confirmed:	1 = 17%
not verifiable:	0
still to check:	4 = 40%

The terms "simple" and " complicated" are defined above, and the term "height n " shows measurements used to predict faults in current operations / 13 /.

The result of complicated tectonics will be explained in more detail:

On thrusts a total of 21 measurements have been carried out. Of these, 7 could be verified, at 6 agreed prediction and startup result. In the other 14 attempts were z. T, no indicators of thrusts can be detected metrologically. The previous interpretation of the measurement results suggests that the angle of inclination between the fault and the seam should not exceed 25 to 30 gon.

In the case of disturbance bundles, indicators can only be expected if the denominations are significantly smaller than the filament size. Because of the resulting energy loss, only a limited number of disturbances can be detected; The prerequisite is that the distances between the individual disturbances are not too low.

Leaching presents a particular problem. Due to the irregular interfaces between leaching and seam, combined with a mostly acute angle of inclination, reflections can not return in an unobstructed trajectory, so that hardly any records can be taken.

Since , in these measurements, the disturbance was not the subject of the investigations, known, simple and well-reflecting disturbances were used here in order to investigate the following points more precisely:

- Difference between explosives and detonating cord

- Measurements in seams with a special structure (root soils and mineral deposits)

As a result, a comparison of measurements with different
wave excitation agents revealed no difference that would
be worth mentioning.

In the course of these measurements, a remarkable influence
on the measuring reliability became apparent. This is
especially true when the intermediate seam medium is very
powerful and highly unstable, or splits into several sub-packs
with interposed carbon strips. As a result, an expansion of
Flözwellen was found on the hanging and lying, so that not
only the seam was to be regarded as a waveguide. The
expansion was the greater, the more rooted the sub-rock was /
11 /.

7   Evaluation Of The Measurement Results

The current status of the flözwellenseismischen apron exploration can be summarized in two points:

1. Simple tectonic influences on the mountain can be well located, even if the scrubbed mass is far below the seam thickness and the distance of the disturbance from the measurement display is 160 to 200 times the seam thickness.

2. More complicated tectonic conditions still pose great difficulties in the prediction, the reasons for which have already been explained.

8　Future Development

The loss of resolution associated with envelope formation prevents the detection of clutter. To remedy this situation, the reflected Flözwellenzug must be attributed to a needle pulse. Theoretical investigations confirm this approach.

Since flute waves are subject to absorption - the higher the frequency, the stronger the absorption - and the loss of energy in reflection occurs, research in this area still needs to be done in order to use the insights gained in data processing. Remedies cause a spatial distribution of energy and thus complicate the indication of a possible disorder. This shows that complex geological structures have to be further investigated by model seismic investigations. This requirement is underlined by the high costs of underground measurement, which depending on the length of the track to be measured amount to 50,000 DM to 150,000 DM /11/ /2/.

9 Summary

With the increasing automation of mining, the underground mining forecourt mining is becoming increasingly important. In order to determine the tectonic content of the apron, a weather-protected measuring system was developed, which detects the position of disturbances by seismic channel waves.

There are two methods for mapping:

1. The Through-Transmission Measurement provides information about a fault in the seam

2. The reflection measurement provides information about the situation.

A great sense of accuracy can be achieved with combined use of both methods. The analysis of 111 measurements shows that geological conditions made the accuracy of the measurements very good, while in geologically complex structures the location of the disturbance proved to be very difficult.

10   Bibliography

/1/   Hudewenz, D., Düllmann, H., Die wirtschaftliche

Beurteilung von Störungsdurchörterungen im Abbau

Zeitschrift Glückauf 1979, Nr. 4, S. 143-148

/2/   Arnetzl, H., Klinge, U., Erfahrungen mit der

Flözwellenseismik bei der Vorfelderkundung

Vortrag am 26.11.1981 im Deutschen Bergbaumuseum,

Bochum

/3/   Arnetzl, H., Grundsätzliches über seismische

Untertagemessungen bei Flözwellen im Steinkohlenbergbau

um Flözwellen näher zu beschreiben, bedarf es folgender

Differenzierung:

/4/   Bentrup, F.-K., Seismische Vorfelderkundung zur

Ortung tektonischer Störungen im Steinkohlenbergbau

Zeitschrift Glückauf 1970, Nr. 19, S. 933-938

/5/   Bentrup, F.-K., Flözdurchschallung aus Tiefbohrlöchern

Zeitschrift Glückauf 1971, Nr. 19, S. 685-690

/6/   Arnetzl, H., Klinge, U., Erfahrungen mit der
Flözwellenseismik bei der Vorfelderkundung

Zeitschrift Glückauf 1982. Nr. 13, S. 658-664

/7/   Rüter, H., Reflexionsseismik unter Tage

Zeitschrift Glückauf 1979, Nr. 16, S. 818-820

/8/   Brentrup, F.-K., Flözwellenseismische
Vorfelderkundung

Zeitschrift Glückauf 1979 Nr. 16, S. 820-823

/9/   Millahn, K. O., Flözwellenseismik, Stand und
Entwicklungen

Hausdruck Prakla-Seismos GmbH, Hannover, 1980

/10/   Rüter, H., Anregung und Empfang von Flözwellen des Love-Typs

Hausdruck Prakla-Seismos GmbH, Hannover, 1980

/11/   Brenntrup, F.-K., et all, Flözwellenseismische Vorfelderkundung mit Hilfe digitaler Messwerterfassung

Schlussbericht über das gleichnamige, vom BMFT geförderte, Forschungsvorhaben, Essen, 1982

/12/   Klar, J., A New Firedamp Proof Instrument For The In-Seam Seismic In Coal Mining

Vortrag, 40. Tagung der European Assiciation of Exploration Geophysicists, Dublin, 1978

/13/   Klinge, U., Entwicklung der Vorfelderkundung

Hausdruck Bergbau AG Lippe, Herne 1984

YOUR KNOWLEDGE HAS VALUE

- We will publish your bachelor's and master's thesis, essays and papers

- Your own eBook and book - sold worldwide in all relevant shops

- Earn money with each sale

Upload your text at www.GRIN.com and publish for free